THE MINES AND MINERALS OF CHESTER COUNTY, PENNSYLVANIA

COLOR IMAGE SUPPLEMENT

Ronald A. Sloto

COPYRIGHT 2016

By Ronald A. Sloto

Cover photograph: Pyromorphite from the Phoenixville mines, Schuylkill Township, Chester County, Pennsylvania, 7.7 cm. Carnegie Museum of Natural History collection CM-10407 (Jefferis 1697). Acquired by William Jefferis in June 1852. Photograph by Debra L. Wilson.

FIGURES

Page

1. Almandine from the Avondale quarry, Avondale 1
2. Rutile from the Atglen area 1
3. Almandine from the Avondale quarry, Avondale 1
4. Rutile V-twin from the Atglen area 1
5. Linarite from the Charlestown mine, Charlestown Township 1
6. Pyrite on calcite from the Charlestown mine, Charlestown Township 2
7. Pyrite on calcite from the Charlestown mine, Charlestown Township 2
8. Drawings of calcite crystals from Wheatley's mines 2
9. Calcite from the Wheatley mine, Schuylkill Township 3
10. Quartz, var. amethyst, from the Charles Carter farm, East Bradford Township 3
11. Kyanite from Copes Bridge, East Bradford Township 3
12. Kyanite from Copes Bridge, East Bradford Township 3
13. Quartz, var. smoky, from the Entrikin farm, East Bradford Township 3
14. Quartz, var. amethyst, from the S. Entrikin farm, East Bradford Township 4
15. Kyanite from East Bradford Township 4
16. Zircon from the Hillsdale paper mill locality, East Bradford Township 4
17. Kyanite from Copes Bridge, East Bradford Township 4
18. Zircon from the Hillsdale paper mill locality, East Bradford Township 4
19. Quartz, var. amethyst, from the Gibbons farm, East Bradford Township 5
20. Ankerite from East Caln Township 5
21. Pyrite, ankerite, and calcite from East Caln Township 5
22. Goethite from the Steitler mine, East Pikeland Township 5
23. Beryl from the Pugh farm, East Nottingham Township 5
24. Wavellite from General Trimble's mine, East Whiteland Township 6
25. Turquoise from General Trimble's mine, East Whiteland Township 7
26. Gibbsite from General Trimble's mine, East Whiteland Township 7
27. Wavellite from General Trimble's mine, East Whiteland Township 7
28. Turquoise from General Trimble's mine, East Whiteland Township 7
29. Wavellite from General Trimble's mine, East Whiteland Township 7
30. Aragonite from the Knickerbocker quarry, East Whiteland Township 8
31. Quartz, var. smoky, collected during construction of the Great Valley Corporate Center, East Whiteland Township 8
32. Pyrolucite from the Knickerbocker landfill property, East Whiteland Township 8
33. Limonite, pseudomorph after pyrite, from the Great Valley Corporate Center, East Whiteland Township 8
34. Quartz, var. smoky, collected during construction of the Great Valley Corporate Center, East Whiteland Township 9
35. Doubly terminated quartz crystal collected during construction of the Great Valley Corporate Center, East Whiteland Township 9
36. Goethite from the Knickerbocker landfill property, East Whiteland Township 9
37. Almandine from the Elk Creek prospect, Elk Township 10
38. Schorl from the Elk Creek prospect, Elk Township 10
39. Beryl from the Elk Creek prospect. Elk Township 10
40. Feldspar from the Elk Creek prospect, Elk Township 11
41. Schorl from the Elk Creek prospect, Elk Township 11
42. Schorl from the Elk Creek prospect, Elk Township 11
43. Zircon from the Elk Creek prospect, Elk Township 11

44.	Chabazite from the Chandler Mill Road locality, Kennett Township	12
45.	Epidote from Kennett Township	12
46.	Epidote on quartz from Kennett Township	12
47.	Quartz and epidote from the Balderson farm, Kennett Township	12
48.	Chabazite from Kennett Township	12
49.	Quartz, var. smoky, from the Dickey farm, Lower Oxford Township	13
50.	Dolomite from Nevin's quarry, London Britain Township	13
51.	Aragonite from Nevin's quarry, London Britain Township	13
52.	Dravite from Nevin's quarry, London Britain Township	13
53.	Calcite from London Grove Township	13
54.	Tremolite from London Grove Township	14
55.	Dravite (brown tourmaline) from Jackson's quarry, London Grove Township	14
56.	Quartz from London Grove Township	14
57.	Apatite from Dr. Allison's farm London Grove Township	14
58.	Actinolite, var. mountain leather, from Hugh's quarry, London Grove Township	14
59.	Tremolite from London Grove Township	15
60.	Dravite from London Grove Township	15
61.	Kyanite from London Grove Township	15
62.	Corundum from Newlin Township	16
63.	Beryl from Newlin Township	16
64.	Diaspore from Corundum Hill, Newlin Township	17
65.	Schorl from Corundum Hill, Newlin Township	17
66.	Corundum from Newlin Township	17
67.	Beryl, var. aquamarine, from Newlin Township	18
68.	Margarite from Corundum Hill, Newlin Township	18
69.	Microcline from Corundum Hill, Newlin Township	18
70.	Euphyllite from Corundum Hill, Newlin Township	18
71.	Diaspore from Newlin Township	18
72.	Corundum crystals in matrix from Corundum Hill, Newlin Township	19
73.	Beryl from Newlin Township	19
74.	Zoisite from Newlin Township	19
75.	Beryl from Corundum Hill, Newlin Township	19
76.	Beryl from Newlin Township	19
77.	Beryl from Newlin Township	19
78.	Almandine from Newlin Township	20
79.	Rutile from the Edwards quarry, Newlin Township	20
80.	Quartz, var. amethyst, from the William Hays farm, Newlin Township	20
81.	Almandine from Newlin Township	20
82.	Almandine from Embreeville, Newlin Township	20
83.	Rutile from the Parksburg area	21
84.	Rutile sixling twin from the Parksburg area	21
85.	Rutile from the Parksburg area,	21
86.	Rutile sixling twin from the Parksburg area	21
87.	Garnet from New Garden Township	22
88.	Muscovite from Pennsbury Township	22
89.	Magnetite inclusions in muscovite from the Swayne quarry, Pennsbury Township	22
90.	Almandine from Pennsbury Township	23
91.	Microcline from the Swayne farm, Pennsbury Township	23
92.	Chondrodite from Harvey's quarry, Pennsbury Township	23
93.	Dolomite from Harvey's quarry, Pennsbury Township	23
94.	Microcline from the Swayne quarry, Pennsbury, Township	23

95.	Ankerite from the Black Rock (Phoenixville) tunnel, Phoenixville	24
96.	Sphalerite with quartz and dolomite from the Black Rock (Phoenixville) tunnel, Phoenixville	24
97.	Quartz and dolomite from the Black Rock (Phoenixville) tunnel, Phoenixville	24
98.	Pyrite on dolomite from the Black Rock (Phoenixville) tunnel, Phoenixville	24
99.	Dolomite from the Black Rock (Phoenixville) tunnel, Phoenixville	24
100.	Quartz on dolomite from the Black Rock (Phoenixville) tunnel, Phoenixville	25
101.	Malachite from the Morris mine, Phoenixville	25
102.	Barite from the Morris mine, Phoenixville	25
103.	Schorl from the Taylor quarry, Pocopson Township	25
104.	Schorl from Pocopson Township	25
105.	Rutilated quartz, var. amethyst, from Pocopson Township	26
106.	Quartz, var. amethyst, from Pocopson Township	26
107.	Quartz, var. amethyst, from Pocopson Township	26
108.	Quartz, var. amethyst, from Pocopson Township	26
109.	Quartz, var. amethyst from Pocopson Township	26
110.	Quartz, var. amethyst from Pocopson Township	27
111.	Rutilated quartz, var. amethyst, from Pocopson Township	27
112.	Quartz, var. amethyst, from the Levi Chrisman farm, Sadsbury Township	27
113.	Quartz, var. amethyst, from Sadsbury Township	27
114.	Anglesite in galena from the Wheatley mine, Schuylkill Township	28
115.	Anglesite in galena from the Wheatley mine, Schuylkill Township	28
116.	Anglesite from the Wheatley mine, Schuylkill Township	28
117.	Anglesite from the Phoenixville mines, Schuylkill Township	28
118.	Anglesite in galena from the Wheatley mine, Schuylkill Township	29
119.	Anglesite in galena from the Phoenixville mines, Schuylkill Township	29
120.	Anglesite on galena from the Wheatley mine, Schuylkill Township	29
121.	Azurite, malachite, and cerussite from the Wheatley mine, Schuylkill Township	29
122.	Azurite from the Wheatley mine, Schuylkill Township	29
123.	Ankerite from the Wheatley mine, Schuylkill Township	30
124.	Barite from the Phoenixville mines, Schuylkill Township	30
125.	Barite from the Phoenixville mines, Schuylkill Township	30
126.	Barite from the Brookdale mine, Schuylkill Township	30
127.	Calcite from the Wheatley mine, Schuylkill Township	30
128.	Calcite on fluorite from the Phoenixville mines, Schuylkill Township	31
129.	Pyrite on calcite from the Wheatley mine, Schuylkill Township	31
130.	Calcite from the Wheatley mine, Schuylkill Township	31
131.	Celestine and calcite from the Wheatley mine, Schuylkill Township	31
132.	Calcite from the Wheatley mine, Schuylkill Township	31
133.	Cerussite from the Wheatley mine, Schuylkill Township	32
134.	Reticulated cerussite from the Wheatley mine, Schuylkill Township	32
135.	Cerussite from the Phoenixville mines, Schuylkill Township	32
136.	Twinned cerussite crystals from the Wheatley mine, Schuylkill Township	32
137.	Cerussite from the Phoenixville mines, Schuylkill Township	32
138.	Cerussite and malachite from the Phoenixville mines, Schuylkill Township	33
139.	Chalcopyrite from the Wheatley mine, Schuylkill Township	33
140.	Copper from the Wheatley mine, Schuylkill Township	33
141.	Dolomite from the Wheatley mine, Schuylkill Township	33
142.	Fluorite coated with calcite from the Wheatley mine, Schuylkill Township	33
143.	Fluorite from the Wheatley mine, Schuylkill Township	34
144.	Fluorite and barite from the Wheatley mine, Schuylkill Township	34
145.	Galena from the Wheatley mine, Schuylkill Township	34

146.	Galena and calcite from the Phoenixville mines, Schuylkill Township	34
147.	Pyromorphite coating galena from the Wheatley mine, Schuylkill Township	34
148.	Galena from the Wheatley mine, Schuylkill Township	35
149.	Hemimorphite from the Wheatley mine, Schuylkill Township	35
150.	Hematite from the Chester County mine, Schuylkill Township	35
151.	Hemimorphite from the Brookdale mine, Schuylkill Township	35
152.	Hemimorphite from the Wheatley mine, Schuylkill Township	35
153.	Linarite from the Wheatley mine, Schuylkill Township	36
154.	Malachite from the Wheatley mine, Schuylkill Township	36
155.	Malachite and azurite from the Wheatley mine, Schuylkill Township	36
156.	Mimetite from the Wheatley mine, Schuylkill Township	36
157.	Mimetite from the Wheatley mine, Schuylkill Township	36
158.	Pyromorphite from the Wheatley mine, Schuylkill Township	37
159.	Pyromorphite from the Phoenixville mines, Schuylkill Township	37
160.	Pyromorphite from the Phoenixville mines, Schuylkill Township	37
161.	Pyromorphite from the Wheatley mine, Schuylkill Township	37
162.	Pyromorphite from the Phoenixville mines, Schuylkill Township	37
163.	Pyromorphite from the Wheatley mine, Schuylkill Township	38
164.	Pyromorphite from the Phoenixville mines, Schuylkill Township	38
165.	Pyromorphite from the Phoenixville mines, Schuylkill Township	38
166.	Pyromorphite from the Brookdale mine, Schuylkill Township	38
167.	Pyromorphite from the Phoenixville mines, Schuylkill Township	38
168.	Quartz from the Phoenixville mines, Schuylkill Township	39
169.	Quartz from the Phoenixville mines, Schuylkill Township	39
170.	Quartz from the Phoenix mine, Schuylkill Township	39
171.	Silver from the Phoenixville mine, Schuylkill Township	39
172.	Sulphur on galena from the Brookdale mine, Schuylkill Township	39
173.	Sphalerite from the Wheatley mine, Schuylkill Township	40
174.	Sphalerite from the Phoenixville mines, Schuylkill Township	40
175.	Sphalerite and calcite on barite from the Wheatley mine, Schuylkill Township	40
176.	Calcite on sphalerite from the Wheatley mine, Schuylkill Township	40
177.	Sphalerite from the Phoenixville mine, Schuylkill Township	40
178.	Vanadinite from the Chester County mine, Schuylkill Township	41
179.	Wulfenite from the Phoenixville mines, Schuylkill Township	41
180.	Wulfenite and pyromorphite from the Wheatley mine, Schuylkill Township	41
181.	Wulfenite from the Phoenixville mines, Schuylkill Township	41
182.	Wulfenite from the Wheatley mine, Schuylkill Township	41
183.	Wulfenite from the Wheatley mine, Schuylkill Township	42
184.	Wulfenite om pyromorphite from Wheatley mine, Schuylkill Township	42
185.	Wulfenite from the Wheatley mine, Schuylkill Township	42
186.	Wulfenite om pyromorphite from Wheatley mine, Schuylkill Township	42
187.	Dumortierite from the Coatesville dumortierite locality, Valley Township	42
188.	Calcite from the W.E. Johnson quarry, Howellville, Tredyffrin Township	42
189.	Calcite from the W.E. Johnson quarry, Howellville, Tredyffrin Township	43
190.	Pyrite from Howellville, Tredyffrin Township	43
191.	Graphite from Byers, Uwchlan Township	43
192.	Schorl from Valley Township	43
193.	Beryl from the Coatesville quarry, Valley Township	43
194.	Phosphuranylite from the Steidler quarry, Valley Township	44
195.	Beryl from the Steidler quarry, Valley Township	44
196.	Columbite from the Steidler quarry, Valley Township	44

197.	Almandine from the Steidler quarry, Valley Township	44
198.	Beryl from the Steidler quarry, Valley Township	45
199.	Columbite from the Steidler quarry, Valley Township	45
200.	Aragonite from West Goshen Township	45
201.	Tremolite from Bailey's quarry, West Marlborough Township	45
202.	Quartz from the Poorhouse quarry, West Bradford Township	46
203.	Microcline, var. chesterlite, from the Poorhouse quarry, West Bradford Township	46
204.	Calcite from the Poorhouse quarry, West Bradford Township	46
205.	Rutile from the Poorhouse quarry, West Bradford Township	46
206.	Microcline, var. chesterlite, from the Poorhouse quarry, West Bradford Township	47
207.	Rutile from the Poorhouse quarry, West Bradford Township	47
208.	Microcline, var. chesterlite, from the Poorhouse quarry, West Bradford Township	47
209.	Quartz from the Poorhouse quarry, West Bradford Township	47
210.	Rutile and microcline, var. chesterlite, from the Poorhouse quarry, West Bradford Township	47
211.	Fluorapatite from the Keystone Trappe Rock quarry, Cornog, Wallace Township	48
212.	Albite, var. adularia, and actinolite, var. byssolite, from the Keystone Trappe Rock quarry, Cornog, Wallace Township	48
213.	Clinozisite from the Keystone Trappe Rock quarry, Cornog, Wallace Township	49
214.	Axinite-(Fe) from the Keystone Trappe Rock quarry, Cornog, Wallace Township	49
215.	Actinolite, var. byssolite, from the Keystone Trappe Rock quarry, Cornog, Wallace Township	49
216.	Prehenite from the Keystone Trappe Rock quarry, Cornog, Wallace Township	49
217.	Pyrrhotite from the Keystone Trappe Rock quarry, Cornog, Wallace Township	49
218.	Prehenite from the Keystone Trappe Rock quarry, Cornog, Wallace Township	50
219.	Reniform pyrite from the Keystone Trappe Rock quarry, Cornog, Wallace Township	50
220.	Saginated quartz from the Keystone Trappe Rock quarry, Cornog, Wallace Township	50
221.	Ancylite from the Keystone Trappe Rock quarry, Cornog, Wallace Township	50
222.	Calcite from the Keystone Trappe Rock quarry, Cornog, Wallace Township	50
223.	Clinozisite from the Keystone Trappe Rock quarry, Cornog, Wallace Township	51
224.	Fluorapatite from the Keystone Trappe Rock quarry, Cornog, Wallace Township	51
225.	Albite, var. adularia, and actinolite, var. byssolite, from the Keystone Trappe Rock quarry, Cornog, Wallace Township	51
226.	Prehenite and actinolite, var. byssolite, from the Keystone Trappe Rock quarry, Cornog, Wallace Township	51
227.	Limonite, psuedomorph after pyrite, on clinozoizite from the Keystone Trappe Rock quarry, Cornog, Wallace Township	51
228.	Chalcopyrite from the French Creek mine, Warwick Township	52
229.	Pyrite on magnetite from the French Creek mine, Warwick Township	52
230.	Garnet from the French Creek mine, Warwick Township	53
231.	Apophyllite from the French Creek mine, Warwick Township	53
232.	Water-clear calcite crystal with pyrite and magnetite from the French Creek mine, Warwick Township	53
233.	Magnetite from the French Creek mine, Warwick Township	53
234.	Azurite and malachite from the Keim (French Creek) mine, Warwick Township	53
235.	Apophyllite from the French Creek mine, Warwick Township	54
236.	Malachite from the French Creek mine, Warwick Township	54
237.	Calcite from the French Creek mine, Warwick Township	54
238.	Orthoclase from the French Creek mine, Warwick Township	54
239.	Natrolite from the French Creek mine, Warwick Township	54
240.	Actinolite, var, byssolite, from the French Creek mine, Warwick Township	55
241.	Gypsum and jarosite from the French Creek mine, Warwick Township	55
242.	Erthryite from the French Creek mine, Warwick Township	55

243.	Twinned octahedral pyrite crystals from the French Creek mine, Warwick Township	55
244.	Diopside from the French Creek mine, Warwick Township	55
245.	Stilpnomelane from the French Creek mine, Warwick Township	56
246.	Sphalerite on actinolite, var, byssolite, with pyrite from the French Creek mine, Warwick Township	56
247.	Pyrite from the French Creek mine, Warwick Township	56
248.	Azurite and malachite from the French Creek mine, Warwick Township	56
249.	Magnetite, pseudomorph after hematite, "hematite rose" from the French Creek mine, Warwick Township	56
250.	Pyrite from the French Creek mine, Warwick Township	57
251.	Andradite from the French Creek mine, Warwick Township	57
252.	Datolite from the French Creek mine, Warwick Township	57
253.	Single sphenoidal chalcopyrite crystal from the French Creek mine, Warwick Township	57
254.	Tourmaline from the French Creek mine, Warwick Township	57
255.	Aragonite from the French Creek mine, Warwick Township	58
256.	Chalcopyrite from the French Creek mine, Warwick Township	58
257.	Apophyllite from the French Creek mine, Warwick Township	58
258.	Pyrite on magnetite from the French Creek mine, Warwick Township	58
259.	Magnetite, pseudomorph after hematite, from the French Creek mine, Warwick Township	58
260.	Pyrite on magnetite from the French Creek mine, Warwick Township	59
261.	Apophyllite from the French Creek mine, Warwick Township	59
262.	Pyrite on apophyllite from the French Creek mine, Warwick Township	59
263.	Pyrite on calcite from the French Creek mine, Warwick Township	59
264.	Garnet from the French Creek mine, Warwick Township	59
265.	Chromian clinochlore, var. kammererite, from the Scott mine, West Nottingham Township	60
266.	Crysotile from the Sparvetta quarry, West Nottingham Township	60
267.	Andradite from the Hopewell mine, Warwick Township	60
268.	Andradite from the Hopewell mine, Warwick Township	60
269.	Magnetite from the Hopewell mine, Warwick Township	60
270.	Andradite from Knauertown (French Creek mine), Warwick Township	60
271.	Clinochlore from Brinton's quarry, Westtown Township	61
272.	Beryl from Brinton's quarry, Westtown Township	61
273.	Vermiculite/hydrobiotite, var. jeffersite, from Brinton's quarry, Westtown Township	61
274.	Clinochlore from Brinton's quarry, Westtown Township	62
275.	"Deweylite" from Brinton's quarry, Westtown Township	62
276.	Aragonite from Brinton's quarry, Westtown Township	62
277.	Chromian clinochlore from Brinton's quarry, Westtown Township	62
278.	Quartz, var. amethyst, from Brinton's quarry, Westtown Township	62
279.	"Deweylite" from Brinton's quarry, Westtown Township	63
280.	Clinochlore from Brinton's quarry, Westtown Township	63
281.	Clinochlore from Brinton's quarry, Westtown Township	63
282.	Schorl from Brinton's quarry, Westtown Township	63
283.	Clinochlore, var. roseite, from Brinton's quarry, Westtown Township	63
284.	Titanite from the Osborne Hill mine, Westtown Township	64
285.	Zoisite from Bath Springs (West Chester Water Works), West Chester	64
286.	Quartz, var. chalcedony, from Willistown Township	64
287.	Spessartine from the Osborne Hill mine, Westtown Township	64
288.	Talc from West Goshen Township	64

THE MINES AND MINERALS OF OF CHESTER COUNTY, PENNSYLVANIA COLOR IMAGE SUPPLEMENT

INTRODUCTION

Chester County has a rich mineral history that spans more than two centuries. George W. Carpenter, writing in the American Journal of Science in 1828, stated: *"Chester county presents to the mineralogist a rich field for investigation. Her limestone, serpentine and gneiss, the predominant rocks of the county, contain inexhaustible beds of interesting minerals, and the numerous quarries every where in operation, greatly facilitate the means of procuring them. These circumstances, with the polite attention manifested towards strangers by the inhabitants of the county, and the singular hospitality which particularly characterizes them, are inducements of the strongest nature for encouraging the mineralogist, to visit this county in preference to almost any section of the country"*

The earliest mention of a mineral from Chester County in the literature was in 1808 by Adam Seybert and Baron Cardeneffez. Seybert, in his *"Catalog of some American Minerals, which are found in different Parts of the United States,"* included serpentine, asbtestus, amianthus, and tremolite from Chester County. Baron Cardeneffez described a *"sulphuret of molybdena found in Chester County, Pennsylvania."*

The mineralogical importance of Chester County was demonstrated during the auction of the Jay Lininger collection of Pennsylvania minerals in May 2005. During the two day auction, 2,240 Pennsylvania mineral specimens were sold, with 607 minerals or 27 percent from Chester County.

The first printing of The Mines and Minerals of Chester County, produced by the traditional printing method, rapidly sold out. To make the book available again, a second printing was produced using "print on demand" technology. However, this technology could not mix gray-scale and color within the file-size limit; therefore, the color insert of mineral photographs included in the first printing was not part of the second printing. This full-color publication makes available those photographs, as well as many others.

Acknowledgments

This supplement would not be possible without the generosity of the institutions and individuals that allowed access to their mineral collections for study and photography. These include (in alphabetical order) Academy of Natural Sciences of Drexel University (with assistance from Douglas Klieger, Ned Gilmore, and Ted Daeschler), Bryn Mawr College (with assistance from Juliet Reed and Maria Luisa Crawford), Bryon Brookmeyer, Carnegie Museum of Natural History (with assistance from Marc Wilson, Debra Wilson, and Richard Souza), Steve Carter, Delaware County Institute of Science (with assistance from Roger Mitchell), Jay Lininger, Jim McEwen, Reading Public Museum (with assistance from Jacquelynn Accetta), Joseph Varady, Union College (with assistance from George Shaw), and the West Chester University Department of Geology and Astronomy (with assistance from LeeAnn Srogi). The author also thanks Doris Biggs, Bryon Brookmeyer, and Debra Wilson for allowing the use of their photographs in this book.

The minerals in this supplement are attributed to the collections at the time of photography. Since then, some of the mineral specimens have found new owners. The Lininger and Varady collections have been dispersed to new owners through auctions. The Brookmeyer collection was acquired by the Carnegie Museum of Natural History in 2014.

Figure 1. Almandine from the Avondale quarry, Avondale, 4.6 cm. Bryn Mawr College collection Rand 6750.

Figure 2. Rutile from the Atglen area, 1.6 cm. Ron Sloto collection 3036A.

Figure 3. Almandine from the Avondale quarry, Avondale, 6 cm. Bryn Mawr College collection Rand 6725.

Figure 4. Rutile V-twin from the Atglen area, 2 cm. Ron Sloto collection 3036B.

Figure 5. Linarite from the Charlestown mine, Charlestown Township, 2.5 cm. Ron Sloto collection 3127.

The Minerals of Chester County

Figure 6. Pyrite on calcite from the Charlestown mine, Charlestown Township. Union College collection 11.4.1-310 (Wheatley collection).

Figure 7. Pyrite on calcite from the Charlestown mine, Charlestown Township. Union College collection 11.4.1-333 (Wheatley collection).

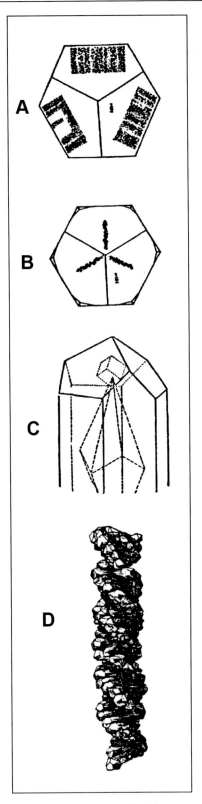

Figure 8. Drawings of calcite crystals from Wheatley's mines. From Smith (1855). Calcite crystals A and B above in Wheatley's collection at Union College are from the Charlestown mine (figs. 6 and 7). A calcite specimen similar to D above is in Wheatley's collection at Union College (fig. 9).

Figure 9. Calcite from the Wheatley mine, Schuylkill Township, 13 cm. Union College collection 11.4.1-4 (Charles Wheatley collection 2524).

Figure 11. Kyanite from Copes Bridge, East Bradford Township, 8.9 cm. West Chester University collection Brinton 256.

Figure 10. Quartz, var. amethyst, from the Charles Carter farm, East Bradford Township, 5.6 cm. Bryn Mawr College collection Vaux 2525.

Figure 12. Kyanite from Copes Bridge, East Bradford Township, 6.2 cm. Ron Sloto collection 3101.

Figure 13. Quartz, var. smoky, from the Entrikin farm, East Bradford Township, 6.2 cm. Carnegie Museum of Natural History collection CM-12828 (Jefferis 7183). Acquired by William Jefferis in 1869.

Figure 14. Quartz, var. amethyst, from the S. Entrikin farm, East Bradford Township, 3.8 cm. Carnegie Museum of Natural History collection CM-1820 (Jefferis 4359).

Figure 15. Kyanite from East Bradford Township, 8 cm. Union College collection.

Figure 16. Zircon from the Hillsdale paper mill locality, East Bradford Township, 2.2 cm. Carnegie Museum of Natural History collection CM-7230 (Jefferis 6680).

Figure 17. Kyanite from Copes Bridge, East Bradford Township, 7 cm. Ron Sloto collection 3340.

Figure 18. Zircon from the Hillsdale paper mill locality, East Bradford Township, 1.2 cm. Bryn Mawr College collection Rand 7032. Collected in 1901.

Figure 19. Quartz, var. amethyst, from the Gibbons farm, East Bradford Township, 4 cm. Carnegie Museum of Natural History collection CM-1828 (Jefferis 815).

Figure 20. Ankerite from East Caln Township, 5 cm. Bryn Mawr College collection Rand 1110.

Figure 21. Pyrite, ankerite, and calcite from East Caln Township. Bryn Mawr College collection Rand 5845.

Figure 22. Goethite from the Steitler mine, East Pikeland Township, 10.5 cm. Ron Sloto collection 2643.

Figure 23. Beryl from the Pugh farm, East Nottingham Township, 6.7 cm. Bryn Mawr College collection Rand 6605.

GENERAL TRIMBLE'S MINE, EAST WHITELAND TOWNSHIP

In 1850, an iron mine was opened on the farm formerly owned by General Trimble, who severed in the Revolutionary War. The mine was worked for a few years and then abandoned. In the 1850s, J. Trimble manufactured polishing powder from the ore. Several shafts were sunk in search of iron ore. In 1856, a horizontal vein of wavellite in stalactites and radiated and crystallized forms was found 10 feet below the land surface. Mineralogists and mineral collectors flocked to the site for specimens. Excellent wavellite specimens also were produced from the 80-foot level. Planerite-turquoise was found in abundance, but was thrown on the dumps as it was considered of no value. About 1880, the mine was reopened, and a 50-foot deep shaft was sunk to the old workings. Several car loads of ore were mined, but the phosphorous content was too high. During this time many fine wavellite specimens were recovered.

This locality was called Steamboat in the early mineralogical literature because it was close to the Steamboat Tavern that once stood on the boundary of East and West Whiteland Townships along the Lancaster Turnpike, now U.S. Business Route 30. When the Pennsylvania Railroad opened passenger service to the settlement around the tavern, the railroad station took its name from the tavern whose sign depicted a steamboat on a body of water. The area was called Steamboat until a post office was established in 1869, and it was renamed Glenlock after the name of a Scottish estate located nearby.

Figure 24. Wavellite from General Trimble's mine, East Whiteland Township, 28 cm. Steve Carter collection. This is probably the finest wavellite specimen from the Trimble mine.

Figure 25. Turquoise from General Trimble's mine, East Whiteland Township, 6.4 cm. Steve Carter collection.

Figure 26. Gibbsite from General Trimble's mine, East Whiteland Township, 9 cm. Bryn Mawr College collection Roedder 23-981.

Figure 27. Wavellite from General Trimble's mine, East Whiteland Township, 11.2 cm. Bryn Mawr College collection Vaux 7831. This specimen was purchased by Vaux from the A.E. Foote Mineral Company in 1894.

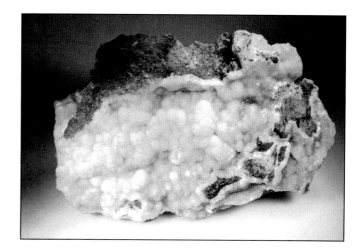

Figure 28. Turquoise from General Trimble's mine, East Whiteland Township, 7 cm. Bryn Mawr College collection Rand 8521.

Figure 29. Wavellite from General Trimble's mine, East Whiteland Township, 7 cm. Bryn Mawr College collection Roedder 980.

EAST WHITELAND TOWNSHIP

Figure 30. Aragonite from the Knickerbocker quarry, East Whiteland Township, 9 cm. Bryn Mawr College collection, Rand 1201.

Figure 31. Quartz, var. smoky, collected during construction of the Great Valley Corporate Center, East Whiteland Township, 6.2 cm. Ron Sloto collection 416B.

Figure 32. Pyrolucite from the Knickerbocker landfill property, East Whiteland Township, 1 cm. Ron Sloto collection 2750.

Figure 33. Limonite, pseudomorph after pyrite, from the Great Valley Corporate Center, East Whiteland Township, 4.5 cm. Ron Sloto collection 206A. It is the largest limonite pseudomorph collected during construction of the Great Valley Corporate Center.

Figure 34. Quartz, var. smoky, collected during construction of the Great Valley Corporate Center, East Whiteland Township, 2.6 cm. Ron Sloto collection 416C.

Figure 35. Doubly terminated quartz crystal collected during construction of the Great Valley Corporate Center, East Whiteland Township, 3.2 cm. Ron Sloto collection 740.

Figure 36. Goethite from the Knickerbocker landfill property, East Whiteland Township, 16 cm. Ron Sloto collection 1120.

ELK CREEK PROSPECT, ELK TOWNSHIP

Figure 37. Almandine from the Elk Creek prospect, Elk Township. Bryon Brookmeyer collection.

Figure 38. Schorl from the Elk Creek prospect, Elk Township, 1.9 cm. Ron Sloto collection 3001A.

Figure 39. Beryl from the Elk Creek prospect. Elk Township, 10.2 cm. Bryon Brookmeyer collection.

Figure 40. Feldspar from the Elk Creek prospect, Elk Township. Crystal is 5 cm. Bryn Mawr College collection Quickel 5552.

Figure 41. Schorl from the Elk Creek prospect, Elk Township, 15.2 cm. Bryon Brookmeyer collection.

Figure 42. Schorl from the Elk Creek prospect, Elk Township, 1.9 cm. Ron Sloto collection 3001B.

Figure 43. Zircon from the Elk Creek prospect, Elk Township, 1 cm. Bryon Brookmeyer collection.

KENNETT TOWNSHIP

Figure 44. Chabazite from the Chandler Mill Road locality, Kennett Township. Field of view is 4.5 cm. Ron Sloto collection 3031.

Figure 45. Epidote from Kennett Township, 4 cm. Carnegie Museum of Natural History collection CM-7612.1.

Figure 46. Epidote on quartz from Kennett Township, 6.8 cm. Carnegie Museum of Natural History collection CM-7589 (Jefferis 10959). Acquired by William Jefferis in 1888.

Figure 47. Quartz and epidote from the Balderson farm, Kennett Township, 5.5 cm. Bryn Mawr College collection, Rand 2120.

Figure 48. Chabazite from Kennett Township, 4 cm. Ron Sloto collection 1111.

Figure 49. Quartz, var. smoky, from the Dickey farm, Lower Oxford Township, 5.7 cm. Bryn Mawr College collection. Collected in 1899.

Figure 50. Dolomite from Nevin's quarry, London Britain Township, 4.4 cm. Bryn Mawr College collection Rand 5817. Collected in 1854.

Figure 51. Aragonite from Nevin's quarry, London Britain Township. Bryn Mawr College collection, Rand 5920.

Figure 52. Dravite from Nevin's quarry, London Britain Township. Crystal is 2 cm. Bryon Brookmeyer collection.

Figure 53. Calcite from London Grove Township, 8.9 cm. Carnegie Museum of Natural History collection CM-4179 (Jefferis 5422).

Figure 54. Tremolite from London Grove Township, 9.5 cm. Ron Sloto collection 2634.

Figure 55. Dravite (brown tourmaline) from Jackson's quarry, London Grove Township. Crystals to 6 mm. Bryn Mawr College collection Rand 7432. Collected in 1854.

Figure 56. Quartz from London Grove Township, 10 cm. Carnegie Museum of Natural History collection, CM-1709 (Jefferis 806).

Figure 57. Apatite from Dr. Allison's farm, London Grove Township. Academy of Natural Sciences of Drexel University Seybert collection 238.

Figure 58. Actinolite, var. mountain leather, from Hugh's quarry, London Grove Township, 5.7 cm. Bryn Mawr College collection Rand 7889.

Figure 59. Tremolite from London Grove Township, 10 cm. Ron Sloto collection 2751.

Figure 60. Dravite from London Grove Township. Crystals to 5 mm. Carnegie Museum of Natural History collection CM-8027 (Jefferis 134).

Figure 61. Kyanite from London Grove Township, 7.2 cm. Carnegie Museum of Natural History collection CM-7478 (Jefferis 8898). Acquired by William Jefferis in 1876.

CORUNDUM HILL AND BERYL HILL, NEWLIN TOWNSHIP

Corundum Hill is in the southeastern corner of Newlin Township in "the barrens," an area of sparse vegetation underlain by serpentinite. Corundum occurs as crystals and masses in pegmatites in the serpentinite. Corundum was first found as large masses lying on the surface; these were a nuisance to farmers plowing their fields. Smaller boulders of corundum were used in stone fences. The farmers tried to drill holes in the larger masses for blasting, but they were too hard. They finally resorted to digging holes and burying the boulders deep enough so that they would not be struck by the plow. Some of the corundum masses were quite large. One mass was estimated to weigh 150 tons and was exposed by trenching. This large mass of corundum attracted much attention and was visited by many prominent mineralogists of the day. Fredrick A. Genth visited the corundum mines in 1872 and described the mass as nearly solid brownish gray granular corundum 30 feet long, 5 to 10 feet wide, and 15 feet thick.

Figure 62. Corundum coated with margarite from Newlin Township, 7.5 cm. Carnegie Museum of Natural History collection CM-12140 (Jefferis 11602). Acquired by William Jefferis in 1893.

The Patterson mine was the first corundum mine in the area. Around 1836, shortly after learning of the discovery of corundum in America, an English plate glass manufacturing firm that used corundum to grind glass sent an agent to Yale College. At Yale, the agent found a 5-pound corundum crystal from Corundum Hill sent by Lewis White Williams. The agent called on Williams, and at his suggestion, Williams and a New York man named Platt purchased the property in Newlin Township where the crystal was found. An old log house stood on the property, and in front of this house was a well. Corundum was discovered on the property while the well was being dug. The well was abandoned when the diggers struck a hard mass of corundum. In 1844, Williams and Platt pumped out the well and deepened it to 50 feet, thus opening the first corundum mine in Chester County.

Figure 63. Beryl from Newlin Township, 10 cm. Carnegie Museum of Natural History collection CM-6497 (Jefferis 8287).

Diaspore crystals from the corundum mines received considerable attention. Diaspore was found attached to a large mass of corundum on the Lesley farm in 1866. Mineralogists flocked to the site to collect specimens.

In 1901, Theoore Rand became the first to refer in print to the corundum area near Unionville as "Corundum Hill." However, local collectors referred to it as such in the 1800s. Minerals from this locality are labeled as Corundum Hill, Beryl Hill, Unionville, and Point Prospect.

Beryl Hill is northeast of Corundum Hill. This locality is named for the beryl crystals found in two pegmatite dikes in serpentinite. Thomas Seal, in a letter written to Benjamin Silliman in 1821, related his discovery 18 months earlier of what later became known as Beryl Hill in the barrens of "New Linn Township." He reported that "the mineral is scattered over the surface, and for the most part in irregular pieces, yet some tolerable crystals are found, from a few grains to 20 lbs. weight." Genth reported that one crystal from Beryl Hill weighed 51 pounds.

Figure 64. Diaspore from Corundum Hill, Newlin Township, 2.5 cm. Steve Carter collection.

Figure 65. Schorl from Corundum Hill, Newlin Township, 4 cm. Ron Sloto collection 2395.

Figure 66. Corundum from Newlin Township. Carnegie Museum of Natural History collection. CM-2527 (Jefferis 4314). A, 4.8 cm. B, 4.2 cm. C, 4 cm.

The Minerals of Chester County

Figure 67. Beryl, var. aquamarine, from Newlin Township, 8 cm. Carnegie Museum of Natural History collection CM-6443 (Jefferis 9438).

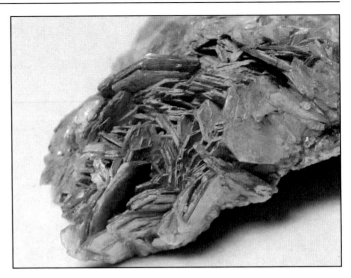

Figure 68. Margarite from Corundum Hill, Newlin Township. Field of view is 4.5 cm. Ron Sloto collection 2612.

Figure 69. Microcline from Corundum Hill, Newlin Township, 7.6 cm. Steve Carter collection.

Figure 70. Euphyllite from Corundum Hill, Newlin Township, 5.8 cm. Ron Sloto collection 2614.

Figure 71. Diaspore from Newlin Township, 4.7 cm. Academy of Natural Sciences of Drexel University Vaux collection 11038.

Figure 72. Corundum crystals in matrix from Corundum Hill, Newlin Township, 6.4 cm. Ron Sloto collection 4.

Figure 74. Zoisite from Newlin Township, 5.5 cm. Bryn Mawr College collection Vaux 5921.

Figure 76. Beryl from Newlin Township, 5.6 cm. Carnegie Museum of Natural History collection CM-6417 (Jefferis 5497).

Figure 77. Beryl from Newlin Township, 3.8 cm. Carnegie Museum of Natural History collection CM-6443 (Jefferis 9438).

Figure 73. Beryl from Newlin Township, 4.9 cm. Carnegie Museum of Natural History collection CM-6474 (Jefferis 9439). Acquired by William Jefferis in 1876.

Figure 75. Beryl from Newlin Township, 6.5 centimeters. Carnegie Museum of Natural History collection CM-6452 (Jefferis 9442).

The Minerals of Chester County

Figure 78. Almandine from Newlin Township, 1.9 cm. Carnegie Museum of Natural History collection CM-6678 (Jefferis 3108).

Figure 80. Quartz, var. amethyst, from the William Hays farm, Newlin Township, 4.4 cm. Carnegie Museum of Natural History collection CM-1831 (Jefferis 10453). Acquired by William Jefferis in 1871.

Figure 79. Rutile from the Edwards quarry, Newlin Township, 1.8 cm. Carnegie Museum of Natural History collection CM-3164 (Jefferis 9458). Acquired by William Jefferis in 1871.

Figure 81. Almandine from Newlin Township, 2.3 cm. Carnegie Museum of Natural History collection CM-6922 (Jefferis 6521). Acquired by William Jefferis in 1870.

Figure 82. Almandine from Embreeville, Newlin Township, 1.3 cm. Ron Sloto collection 2638.

RUTILE FROM PARKSBURG

Figure 83. Rutile from the Parksburg area. Bryon Brookmeyer collection. Jeff Scovil photograph. Courtesy of Brian Brookmeyer.

Figure 84. Rutile sixling twin from the Parksburg area, 5.5 cm. Carnegie Museum of Natural History collection CM-21515.

Figure 85. Rutile from the Parksburg area, 6 cm. Delaware County Institute of Science collection.

Figure 86. Rutile sixling twin from the Parksburg area, 2.8 cm. Bryn Mawr College collection Roedder 908.

Rutile

Rutile (titanium dioxide) is found in Chester Valley between Parksburg and Atglen. Rutile crystals are often twinned, producing V twins and six-sided cyclical twins known as "sixlings."

Rutile, locally called money stone, was collected from plowed fields and sold to mineral collectors and the dental trade. The rutile was worth 10 to 12 cents per ounce. It was used in the manufacture of false teeth.

Figure 87. Garnet from New Garden Township, 5 cm. Bryon Brookmeyer collection.

Figure 88. Muscovite from Pennsbury Township, 14.5 cm. Ron Sloto collection 2626.

Figure 89. Magnetite inclusions in muscovite from the Swayne quarry, Pennsbury Township, 7.5 cm. Ron Sloto collection 2688.

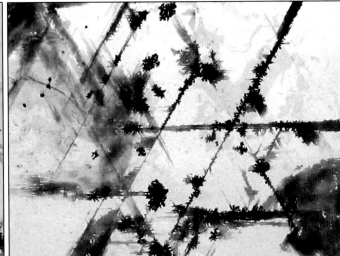

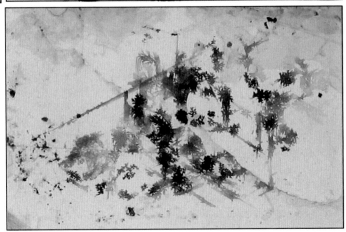

Figure 90. Almandine from Pennsbury Township, 9.4 cm. Carnegie Museum of Natural History collection CM-6747 (Jefferis 676).

Figure 91. Microcline from the Swyane farm, Pennsbury Township, 3.5 cm. Bryn Mawr College collection Rand 6133.

Figure 92. Chondrodite from Harvey's quarry, Pennsbury Township, 6.6 cm. This is the only locality in Pennsylvania where chondrodite occurs. Ron Sloto collection 422.

Figure 93. Dolomite from Harvey's quarry, Pennsbury Township. Field of view is 4 cm. Ron Sloto collection 2000.

Figure 94. Microcline from the Swayne quarry, Pennsbury Township, 5.7 cm. Bryn Mawr College collection, Rand 6113.

BLACK ROCK (PHOENIXVILLE) TUNNEL

Figure 95. Ankerite from the Black Rock (Phoenixville) tunnel, Phoenixville, 7.6 cm. Ron Sloto collection 3448.

Figure 96. Sphalerite with quartz and dolomite from the Black Rock (Phoenixville) tunnel, Phoenixville. Sphalerite crystal is 9 mm. Carnegie Museum of Natural History collection ANSP-24699.

Figure 97. Quartz and dolomite from the Black Rock (Phoenixville) tunnel, Phoenixville, 4.8 cm. Ron Sloto collection 2664.

Figure 98. Pyrite on dolomite from the Black Rock (Phoenixville) tunnel, Phoenixville, 7 cm. Carnegie Museum of Natural History collection CM-4293 (Jefferis 5330).

Figure 99. Dolomite from the Black Rock (Phoenixville) tunnel, Phoenixville, 9 cm. West Chester University collection Gay collection.

Figure 100. Quartz on dolomite from the Black Rock (Phoenixville) tunnel, Phoenixville, 7.6 cm. Bryon Brookmeyer collection.

Figure 101. Barite from the Morris mine, Phoenixville, 2.5 cm. Ron Sloto collection 3594.

Figure 102. Malachite from the Morris mine, Phoenixville, 7.5 cm. Ron Sloto collection 3364.

Figure 103. Schorl from the Talyor quarry, East Marlboro Township, 2.5 cm. Bryn Mawr College collection Rand 7410.

Figure 104. Schorl from Pocopson Township, 4 cm. Ron Sloto collection 3056.

AMETHYST FROM POCOPSON TOWNSHIP

Figure 105. Rutilated quartz, var. amethyst, from Pocopson Township, 3.5 cm. Ron Sloto collection 3042A.

Figure 106. (left) Quartz, var. amethyst, from Pocopson Township, 2 cm, Ron Sloto collection 561A. Collected in 1984.

Figure 107. (right) Quartz, var. amethyst, from Pocopson Township, 2 cm. Ron Sloto collection 561B.

Figure 108. (left) Quartz, var. amethyst, from Pocopson Township, 1.8 cm. Ron Sloto collection 561D.

Figure 109. (right) Quartz, var. amethyst from Pocopson Township, 2.5 cm. It is rare to find matrix attached to the crystals. Ron Sloto collection 561C.

Figure 110. Quartz, var. amethyst from Pocopson Township, 2.1 cm. Ron Sloto collection 561E.

Figure 111. Rutilated quartz, var. amethyst, from Pocopson Township, 4.8 cm. Ron Sloto collection 3042B.

AMETHYST FROM SADSBURY TOWNSHIP

Figure 112. Quartz, var. amethyst, from the Levi Chrisman farm, Sadsbury Township, 1.5 cm. Bryn Mawr College collection Rand 4860. Collected in 1894.

Figure 113. Quartz, var. amethyst, from Sadsbury Township, 4.3 cm. Carnegie Museum of Natural History collection CM-1887 (Jefferis 809).

PHOENIXVILLE DISTRICT LEAD MINES
SCHUYLKILL TOWNSHIP

Figure 114. Anglesite in galena from the Wheatley mine, Schuylkill Township, 10 cm. Anglesite crystal is 2.5 cm. Carnegie Museum of Natural History collection CM-19377.

Figure 115. Anglesite in galena from the Wheatley mine, Schuylkill Township. Crystals are 1 cm. Carnegie Museum of Natural History collection CM-26835.

Figure 116. Anglesite from the Wheatley mine, Schuylkill Township, 4 cm. Union College collection 25.7.1-8 (Charles Wheatley collection 2187).

Figure 117. Anglesite from the Phoenixville mines, Schuylkill Township, 7 mm. Bryn Mawr College collection Vaux 8292.

Figure 118. Anglesite in galena from the Wheatley mine, Schuylkill Township. Carnegie Museum of Natural History collection CM-27625. Photo by Debra K. Wilson.

Figure 119. Anglesite in galena from the Phoenixville mines, Schuylkill Township. Anglesite crystals to 2.5 cm. Bryon Brookmeyer collection. Former H.D. Rogers specimen.

Figure 120. Anglesite on galena from the Wheatley mine, Schuylkill Township. Anglesite crystal is 2.5 cm. Union College collection 25.7.1-10 (Charles Wheatley collection).

Figure 121. Azurite, malachite, and cerussite from the Wheatley mine, Schuylkill Township. Union College collection 11.2.2-42 (Charles Wheatley collection 3779).

Figure 122. Azurite from the Wheatley mine, Schuylkill Township. Crystals to 8 mm. Union College collection 11.2.2-41 (Charles Wheatley collection 3753).

Figure 123. Ankerite from the Wheatley mine, Schuylkill Township. Carnegie Museum of Natural History collection ANSP-6100.

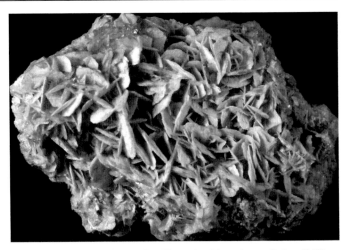

Figure 124. Barite from the Phoenixville mines, Schuylkill Township, 6.5 cm. Jay Lininger collection.

Figure 125. Barite from the Phoenixville mines, Schuylkill Township. Jay Lininger collection.

Figure 126. Barite from the Brookdale mine, Schuylkill Township, 5 cm. Steve Carter collection.

Figure 127. Calcite from the Wheatley mine, Schuylkill Township. Calcite crystals to 1.2 cm. Union College collection 11.4.1-85 (Charles Wheatley collection 2579).

Figure 128. Calcite on fluorite from the Phoenixville mines, Schuylkill Township, 7.5 cm. Bryon Brookmeyer collection.

Figure 129. Pyrite on calcite from the Wheatley mine, Schuylkill Township, 6.5 cm. Union College collection 11.4.1-346 (Charles Wheatley collection 4176).

Figure 130. Calcite from the Wheatley mine, Schuylkill Township, 4 cm. Carnegie Museum of Natural History collection CM-3863 (Jefferis 4348).

Figure 131. Celestine and calcite from the Wheatley mine, Schuylkill Township. Union College collection 25.4.12-40 (Charles Wheatley collection 4699).

Figure 132. Calcite from the Wheatley mine, Schuylkill Township. Calcite crystals to 4 cm. Carnegie Museum of Natural History collection ANSP-38.

Figure 133. Cerussite from the Wheatley mine, Schuylkill Township. Crystals to 7 mm. Bryn Mawr College collection Vaux 411.

Figure 134. Reticulated cerussite from the Wheatley mine, Schuylkill Township. Union College collection. 11.9.1-36 (Charles Wheatley collection 3697).

Figure 135. Cerussite from the Wheatley mine, Schuylkill Township, 6 cm. Union College collection 11.9.1-28.

Figure 136. Twinned cerussite crystals from the Phoenixville mines, Schuylkill Township, 3.2 cm. Bryon Brookmeyer collection.

Figure 137. Cerussite from the Phoenixville mines, Schuylkill Township, 5 cm. Bryon Brookmeyer collection.

Figure 138. Cerussite and malachite from the Phoenixville mines, Schuylkill Township. Carnegie Museum of Natural History collection ANSP-25727.

Figure 139. Chalcopyrite from the Wheatley mine, Schuylkill Township, 4 cm. Bryn Mawr College collection Arndt 1137.

Figure 140. Copper from the Wheatley mine, Schuylkill Township, 2.5 cm. Bryn Mawr College collection Heyl 1680.

Figure 141. Dolomite from the Wheatley mine, Schuylkill Township, 8.5 cm. Carnegie Museum of Natural History collection CM-4320 (Jefferis 2442). Acquired by William Jefferis in October 1852.

Figure 142. Fluorite coated with calcite from the Wheatley mine, Schuylkill Township. Fluorite crystals to 1.5 cm. Jay Lininger collection.

Figure 143. Fluorite from the Wheatley mine, Schuylkill Township, 13 cm. Joseph Varaday collection.

Figure 144. Fluorite and barite from the Wheatley mine, Schuylkill Township, 5 cm. Bryon Brookmeyer collection.

Figure 145. Galena from the Wheatley mine, Schuylkill Township, 4.5 cm. Bryn Mawr College collection Vaux 746.

Figure 146. Galena and calcite from the Phoenixville mines, Schuylkill Township, 12 cm. Bryon Brookmeyer collection.

Figure 147. Pyromorphite coating galena from the Wheatley mine, Schuylkill Township, 13 cm. Carnegie Museum of Natural History collection CM-27627.

Figure 148. Galena from the Wheatley mine, Schuylkill Township, 15 cm. Union College collection 3.6.3-92 (Charles Wheatley collection 197).

Figure 149. Hemimorphite from the Wheatley mine, Schuylkill Township. Jay Lininger collection.

Figure 150. Hematite from the Chester County mine, Schuylkill Township, 8.5 cm. Bryn Mawr College collection Rand 5519.

Figure 151. Hemimorphite from the Brookdale mine, Schuylkill Township, 7 cm. Carnegie Museum of Natural History collection CM-7837 (Jefferis 1970). Acquired by William Jefferis in 1854.

Figure 152. Hemimorphite from the Wheatley mine, Schuylkill Township, 8 cm. Union College collection 11.7.5-52 (Charles Wheatley collection 1805).

Figure 153. Linarite from the Wheatley mine, Schuylkill Township, 1.7 cm. Bryn Mawr College collection Heyl 2650.

Figure 154. Malachite from the Wheatley mine, Schuylkill Township, 8 cm. Jay Lininger collection.

Figure 156. Mimetite from the Wheatley mine, Schuylkill Township. Crystals to 2 mm. Bryn Mawr College collection Vaux 7399A.

Figure 155. Malachite and azurite from the Wheatley mine, Schuylkill Township, 7 cm. Union College collection 11.2.1-51 (Charles Wheatley collection 3757).

Figure 157. Mimetite from the Wheatley mine, Schuylkill Township, 4.2 cm. Bryn Mawr College collection Vaux 7399B.

Figure 158. Pyromorphite from the Wheatley mine, Schuylkill Township, 17 cm. Union College collection 22.2.5-12 (Charles Wheatley collection).

Figure 159. Pyromorphite from the Phoenixville mines, Schuylkill Township, 6.6 cm. Carnegie Museum of Natural History collection CM-10407 (Jefferis 1697). Acquired by William Jefferis in September 1852.

Figure 160. Pyromorphite from the Phoenixville mines, Schuylkill Township. Pyromorphite crystals to 7 mm. Bryon Brookmeyer collection.

Figure 161. Pyromorphite from the Wheatley mine, Schuylkill Township, 8 cm. Union College collection 22.2.5-23 (Charles Wheatley collection).

Figure 162. Pyromorphite from from the Phoenixville mines, Schuylkill Township. Academy of Natural Sciences of Drexel University Vaux collection 14024.

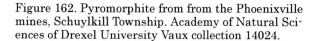

Figure 163. Pyromorphite from the Wheatley mine, Schuylkill Township, 8.5 cm. Union College collection 22.2.5-45 (Wheatley collection).

Figure 165. Pyromorphite from the Phoenixville mines, Schuylkill Township, 9 cm. Bryon Brookmeyer collection.

Figure 164. Pyromorphite from the Phoenixville mines, Schuylkill Township. Ron Sloto collection 3535. Photo appeared on the cover of the January 1992 Lapidary Journal.

Figure 166. Pyromorphite from the Brookdale mine, Schuylkill Township. Pyromorphite crystals to 3 mm. Carnegie Museum of Natural History collection CM-10460 (Jefferis 2000). Acquired by William Jefferis in 1855.

Figure 167. Pyromorphite from the Phoenixville mines, Schuylkill Township, 4 cm. Bryon Brookmeyer collection.

Figure 168. Quartz from the Phoenixville mines, Schuylkill Township, 8.8 cm. Carnegie Museum of Natural History collection CM-1772 (Jefferis 6653).

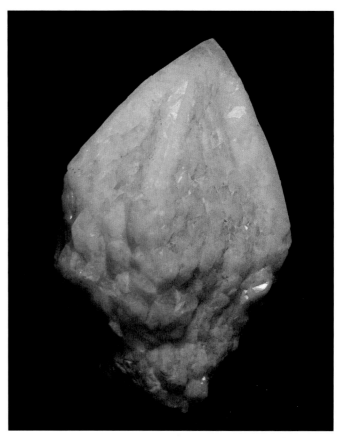

Figure 169. Quartz from the Phoenixville mines, Schuylkill Township, 23 cm. West Chester University collection.

Figure 170. Quartz from the Phoenix mine, Schuylkill Township, 5 cm. Ron Sloto collection 2710.

Figure 171. Silver from the Phoenixville mines, Schuylkill Township, magnified. Jay Lininger collection.

Figure 172. Sulphur on galena from the Brookdale mine, Schuylkill Township, magnified. Carnegie Museum of Natural History collection CM-86 (Jefferis 137). Acquired by William Jefferis in 1859.

Figure 173. Sphalerite from the Wheatley mine, Schuylkill Township, 10 cm. Carnegie Museum of Natural History collection CM-19127.

Figure 174. Sphalerite from the Phoenixville mines, Schuylkill Township, 5.5 cm. Bryon Brookmeyer collection.

Figure 176. Calcite on sphalerite from the Wheatley mine, Schuylkill Township, 24 cm. Carnegie Museum of Natural History collection CM-27626.

Figure 175. Sphalerite and calcite on barite from the Wheatley mine, Schuylkill Township. Union College collection 3.4.2-31 (Charles Wheatley collection).

Figure 177. Sphalerite from the Phoenixville mines, Schuylkill Township, 16 cm. Academy of Natural Sciences of Drexel University Vaux collection 10321.

Figure 178. Vanadinite from the Chester County mine, Schuylkill Township, magnified. Ron Sloto collection 1301.

Figure 179. Wulfenite from the Phoenixville mines, Schuylkill Township. Wulfenite crystals to 2 mm. Steve Carter collection.

Figure 180. Wulfenite and pyromorphite from the Wheatley mine, Schuylkill Township. Wulfenite crystals to 2 mm. Carnegie Museum of Natural History collection CM-11813 (Jefferis 2821).

Figure 181. Wulfenite from the Phoenixville mines, Schuylkill Township. Wulfenite crystals to 2 mm. Bryon Brookmeyer collection.

Figure 182. Wulfenite from the Wheatley mine, Schuylkill Township. Wulfenite crystals to 4 mm. Bryn Mawr College collection Vaux 8673.

Figure 183 Wulfenite from the Wheatley mine, Schuylkill Township. Bryn Mawr College collection Heyl 3644.

Figure 184. Wulfenite on pyromorphite from the Wheatley mine, Schuylkill Township. Wulfenite crystals to 2 mm. Carnegie Museum of Natural History collection CM-1108 (Jefferis 1810). Acquired by William Jefferis in 1854.

Figure 185. Wulfenite from the Wheatley mine, Schuylkill Township. Wulfenite crystals to 2 mm. Bryn Mawr College collection Vaux 8669.

Figure 186. Wulfenite on pyromorphite from the Wheatley mine, Schuylkill Township. Union College collection 27.2.3-28 (Charles Wheatley collection 3507).

Figure 187. Dumortierite from the Coatesville dumortierite locality, Valley Township, 10.8 cm. Bryon Brookmeyer collection.

Figure 188. Calcite from the W.E. Johnson quarry, Howellville, Tredyffrin Township, 3.7 cm. Ron Sloto collection 2752C.

Figure 189. Calcite from the W.E. Johnson quarry, Howellville, Tredyffrin Township, 3.7 cm. Ron Sloto collection 2752B.

Figure 190. Pyrite from Howellville, Tredyffrin Township. Pyrite crystal is 1.4 cm. Bryn Mawr College collection Vaux 1178.

Figure 191. Graphite from Byers, Uwchlan Township, 14.5 cm. Ron Sloto collection 3299.

Figure 192. Schorl from Valley Township, 2.5 cm. Ron Sloto collection 3343.

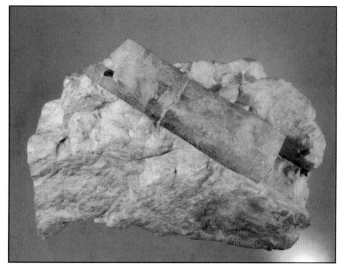

Figure 193. Beryl from the Coatesville quarry, Valley Township, 5 cm. Steve Carter collection.

STEIDLER QUARRY, VALLEY TOWNSHIP

Figure 194. Phosphuranylite from the Steidler quarry, Valley Township, 8.3 cm. Steve Carter collection.

Figure 195. Beryl from the Steidler quarry, Valley Township, 4.4 cm. Steve Carter collection.

Figure 196. Columbite from the Steidler quarry, Valley Township. Columbite crystal is 1 cm. Ron Sloto collection 3088.

Figure 197. Almandine from the Steidler quarry, Valley Township, 4.9 cm. Ron Sloto collection 320.

Figure 198. Beryl from the Steidler quarry, Valley Township, 13.6 cm. Ron Sloto collection 2607.

Figure 199. Columbite from the Steidler quarry, Valley Township, 4.5 cm. Ron Sloto collection 322.

Figure 200. Aragonite from West Goshen Township, 6.4 cm. West Chester University collection.

Figure 201. Tremolite from Bailey's quarry, West Marlborough Township, 8 cm. Bryn Mawr College collection Rand 6446.

POORHOUSE QUARRY, WEST BRADFORD TOWNSHIP

The Poorhouse quarry is northeast of Embreeville. It is the only place in Chester County where the contact between the Cockeysville marble and the Wissahickon Schist is exposed.

The Poorhouse quarry derives its name from its location on the property of the former Chester County Poor House, which was chartered by the Commonwealth of Pennsylvania on February 27, 1798, for those unable to support themselves. The poor house was built in 1800 on the 350-acre Harlan farm. It was a working farm and a productive business enterprise that sold diverse commodities, such as weaving and lime.

The Poorhouse quarry was opened in the 1790s. It supplied marble for lime and building stone for construction of buildings on the poorhouse property. In the 1870s, the quarry was operated as a source of agricultural lime produced by burning marble in kilns on the property.

The Poorhouse quarry is most famous for chesterlite, a variety of microcline. Minerals occur in gash-like crevices in the marble, which are lined with crystals of chesterlite, calcite, dolomite, rutile, and clear to milky quartz. The crystals are frequently coated with a thin film of limonite, which gives them an iridescent appearance.

Figure 202. Quartz from the Poorhouse quarry, West Bradford Township. Quartz crystal is 2.3 cm. Bryn Mawr College collection Rand 4788.

Figure 203. Microcline, var. chesterlite from the Poorhouse quarry, West Bradford Township, 6.4 cm. Steve Carter collection.

Figure 204. Calcite from the Poorhouse quarry, West Bradford Township, 13.3 cm. Bryon Brookmeyer collection. Collected March 6, 1873.

Figure 205. Rutile from the Poorhouse quarry, West Bradford Township, 6 mm. Carnegie Museum of Natural History collection CM-3134 (Jefferis 481).

Figure 206. Microcline, var. chesterlite from the Poorhouse quarry, West Bradford Township, 7.6 cm. Steve Carter collection.

Figure 207. Rutile from the Poorhouse quarry, West Bradford Township, 8 mm. Carnegie Museum of Natural History collection CM-3186 (Jefferis 481).

Figure 208. Microcline, var. chesterlite, from the Poorhouse quarry, West Bradford Township, 4.9 cm. Carnegie Museum of Natural History collection CM-5412.

Figure 209. Quartz from the Poorhouse quarry, West Bradford Township, 7.2 cm. Carnegie Museum of Natural History collection CM-2167.

Figure 210. Rutile and microcline, var. chesterlite. from the Poorhouse quarry, West Bradford Township, 1.5 cm. Bryon Brookmeyer collection.

The Minerals of Chester County

KEYSTONE TRAPPE ROCK QUARRY, CORNOG, WALLACE TOWNSHIP

Figure 211 Fluorapatite from the Keystone Trappe Rock quarry, Cornog, Wallace Township. Bryon Brookmeyer collection. Jeff Scovil photograph. Courtesy of Bryon Brookmeyer.

Figure 212. Albite, var. adularia, and actinolite, var. byssolite, from the Keystone Trappe Rock quarry, Cornog, Wallace Township. Bryon Brookmeyer collection.

The Keystone Trappe Rock quarry, commonly known as the Cornog quarry, was west of Cornog. The quarry is abandoned and flooded, and the quarry area is surrounded by a chain-link fence topped with barbed wire.

The Cornog quarry was opened in 1907 by Harry T.A. Rhodewalt, who produced crushed stone with a steam-operated crusher. In 1913, the quarry was leased to John B. Galt. Galt and his brothers operated the quarry under the name Keystone Trappe Rock Company until it was abandoned in 1968 and allowed to flood. The quarry attained a size of 500 feet wide, 700 feet long, and 200 feet deep by the time it was abandoned. It was acquired by Aqua Pennsylvania, Inc. as a future water-supply reservoir.

The suite of minerals exposed by the Cornog quarry is similar to Alpine-cleft occurrences where the chemical constituents of the pocket crystals have been derived from the wall rock, and newly formed species chiefly represent minerals common in or chemically related to the adjoining rocks. The newly formed crystals were deposited in the cavities by hydrothermal solutions, most likely associated with the intrusion of the abundant, nearby granitic pegmatite dikes. Some of the prehnite and albite crystals enclose a dense matting of actinolite, giving the crystals a greenish-gray color.

Figure 213. Clinozisite from the Keystone Trappe Rock quarry, Cornog, Wallace Township, magnified. Bryon Brookmeyer collection.

Figure 215. Actinolite, var. byssolite, from the Keystone Trappe Rock quarry, Cornog, Wallace Township, 8.5 cm. Ron Sloto collection 2647.

Figure 216. Prehnite from the Keystone Trappe Rock quarry, Cornog, Wallace Township, 7.8 centimeters. Ron Sloto collection 2650.

Figure 214. Axinite-(Fe) from the Keystone Trappe Rock quarry, Cornog, Wallace Township, 1.9 cm. Ron Sloto collection 2664.

Figure 217. Pyrrhotite from the Keystone Trappe Rock quarry, Cornog, Wallace Township. Biggs collection. Photograph courtesy of Doris Biggs.

Figure 218. Prehnite from the Keystone Trappe Rock quarry, Cornog, Wallace Township, 3 cm. Ron Sloto collection 2652.

Figure 219. Reniform pyrite from the Keystone Trappe Rock quarry, Cornog, Wallace Township, 3.7 cm. Bryon Brookmeyer collection.

Figure 221. Ancylite from the Keystone Trappe Rock quarry, Cornog Wallace Township, magnified. Biggs collection. Julius Weber photograph. Courtesy of Doris Biggs.

Figure 220. Saginated quartz from the Keystone Trappe Rock quarry, Cornog, Wallace Township, 10 cm. The color is caused by included acicular actinolite crystals. Ron Sloto collection 2648.

Figure 222. Calcite from the Keystone Trappe Rock quarry, Cornog, Wallace Township, field of view is 3.5 cm. Ron Sloto collection 1800.

Figure 223. Clinozisite from the Keystone Trappe Rock quarry, Cornog, Wallace Township. Field of view is 2 cm. Ron Sloto collection 631A.

Figure 224. Fluorapatite from the Keystone Trappe Rock quarry, Cornog, Wallace Township, 1.8 cm. Carnegie Museum of Natural History collection CM-29289.

Figure 226. Prehnite, and actinolite, var. byssolite, from the Keystone Trappe Rock quarry, Cornog, Wallace Township. Field of view is 2 cm. Ron Sloto collection 3115.

Figure 225. Albite, var. adularia, and actinolite, var. byssolite, from the Keystone Trappe Rock quarry, Cornog Wallace Township. Bryon Brookmeyer collection.

Figure 227. Limonite, psuedomorph after pyrite, on clinozoizite from the Keystone Trappe Rock quarry, Cornog, Wallace Township, 4.2 cm. Ron Sloto collection 2655.

The Minerals of Chester County

FRENCH CREEK MINE, SAINT PETERS WARWICK TOWNSHIP

The French Creek mine was the most significant iron mine in the French Creek valley and the most economically significant underground mine in Chester County. It produced an estimated 1 million tons of ore during its 84 years of operation. Two ore bodies were mined.

In March 1845, Samuel Crossley, a stage coach driver, started mining iron ore in several shallow pits. Two years later, Crossley sold the entire 105-acre tract to Jonathan Keim. In 1848, Keim sold the mine property to the E. and G. Brooke Iron Company. During the Civil War, the French Creek Mining Company, also called the French Creek Copper Company, leased the mine for copper ore, but was unsuccessful. By May 1864, the mine produced 1,000 tons of iron ore and only 45 tons of copper ore. Only 200 to 300 tons of iron ore and 10 tons of copper ore were sold. The French Creek Copper Company experienced financial trouble due to the difficulty in transporting the ore to market.

It is not known exactly when E. & G. Brooke and the Phoenix Iron Company reopened the mine, but it was probably in full operation by 1876. Kilns for roasting the ore to remove the sulfur were installed at the mine in 1878. The roasted ore was shipped to the E. & G. Brooke iron furnace in Birdsboro and to the nearby Hopewell furnace by 18 four- and six-horse teams. In 1880, the Wilmington and Northern Railroad completed a branch line to the French Creek mine, thereby solving the ore transportation problem. The railroad named the station at its terminus the Saint Peters station.

The French Creek mine closed in 1895 because the E. & G. Brooke Iron Company found it more economical to import iron ore than to mine it. At the time the mine closed in 1895, the no. 2 mine had been developed to the 250-foot level, and the shaft of the no. 1 mine reached the 360-foot level.

After a 17-year lapse, new activity began at the French Creek mine in 1912. E. & G. Brooke reopened and modernized the no. 1 mine. The mine was dewatered, the shaft retimbered, and the mine equipped with electricity and compressed air. The shaft and slope were remodeled into a compound shaft and extended for 100 feet along the footwall, reaching a vertical distance of 485 feet. Production began on June 15, 1914.

With the outbreak of World War I, the demand for iron increased, and a new 43° inclined shaft was started at mine no. 1. The new shaft was sunk 150 yards southeast of the old shaft and was located 100 feet back in the footwall. In 1926, when the no. 1 mine ore body was nearing depletion, E. & G. Brooke hired several consultants to search for a new ore body. A surface geophysical survey and test drilling failed to reveal a new ore body. In 1928, the no. 1 mine reached the 1,250-foot level where the ore body ended abruptly against a dike of hornblende syenite gneiss. With the ore body depleted, the French Creek mine closed in 1928.

Figure 228. Chalcopyrite from the French Creek mine, Warwick Township, 10 cm. Jay Lininger collection.

Figure 229. Pyrite on magnetite from the French Creek mine, Warwick Township, 8 cm. Bryn Mawr College collection.

Figure 230. Garnet from the French Creek mine, Warwick Township. Field of view is 5 cm. Ron Sloto collection 2150.

Figure 231. Apophyllite from the French Creek mine, Warwick Township, 3.6 cm. Carnegie Museum of Natural History collection.

Figure 232. Water-clear calcite crystal with pyrite and magnetite from the French Creek mine, Warwick Township, 8.3 cm. Steve Carter collection.

Figure 233. Magnetite from the French Creek mine, Warwick Township, 5.1 cm. Ron Sloto collection 3377.

Figure 234. Azurite and malachite from the Keim (French Creek) mine Warwick Township, 10.2 cm. Steve Carter collection.

Figure 235. Apophyllite from the French Creek mine, Warwick Township, 16 cm. Bryon Brookmeyer collection.

Figure 236. Malachite from the French Creek mine, Warwick Township, 6.4 cm. Steve Carter collection.

Figure 237. Calcite from the French Creek mine, Warwick Township, 10 cm. Bryn Mawr College collection Rand 1003.

Figure 238. Orthoclase from the French Creek mine, Warwick Township, 8.5 cm. Bryn Mawr College collection Rand 1319-A. Collected in 1896.

Figure 239. Natrolite from the French Creek mine, Warwick Township, 8 cm. Bryn Mawr College collection.

Figure 240. Actinolite, var. byssolite, from the French Creek mine, Warwick Township, 5.4 cm. Ron Sloto collection 2134.

Figure 241. Gypsum and jarosite from the French Creek mine, Warwick Township. Field of view is 3 cm. Ron Sloto collection 1267.

Figure 242. Erthryite from the French Creek mine, Warwick Township, 5 cm. Ron Sloto collection 1256.

Figure 243. Twinned octahedral pyrite crystals from the French Creek mine, Warwick Township, 1.8 cm. Bryn Mawr College collection. Collectors in the 1800s sometimes mounted single crystals on wooden pedestals.

Figure 244. Diopside from the French Creek mine, Warwick Township, magnified. Ron Sloto collection 2018.

The Minerals of Chester County

Figure 245. Stilpnomelane from the French Creek mine, Warwick Township, 4.1 cm. Ron Sloto collection 291.

Figure 246. Sphalerite on actinolite, var. byssolite, from the French Creek mine, Warwick Township, 5 cm. Sphalerite crystals are 1.9 cm. Bryon Brookmeyer collection.

Figure 247. Pyrite from the French Creek mine, Warwick Township, 5.6 cm. Carnegie Museum of Natural History collection ANSP-24729. Collected by Sam Gordon in 1924.

Figure 248. Azurite and malachite from the French Creek mine, Warwick Township, 15 cm. Steve Carter collection.

Figure 249. Magnetite, pseudomorph after hematite, "hematite rose" from the French Creek mine, Warwick Township, 2.7 cm. Bryn Mawr College collection Rand 5359.

Figure 250. Pyrite from the French Creek mine, Warwick Township, 4 cm. Ron Sloto collection 3161.

Figure 251. Andradite from the French Creek mine, Warwick Township, 9 cm. Jay Lininger collection.

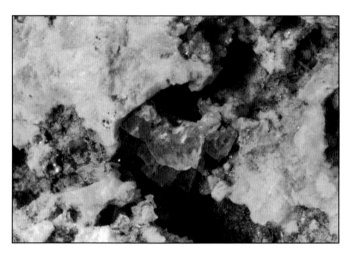

Figure 252. Datolite from the French Creek mine, Warwick Township, magnified. Ron Sloto collection 2470.

Figure 253. Single sphenoidal chalcopyrite crystal from the French Creek mine, Warwick Township, 4.5 cm. Bryon Brookmeyer collection.

Figure 254. Tourmaline from the French Creek mine, Warwick Township, magnified. Union College collection.

Figure 255. Grossular from the Elizabeth (French Creek) mine, Warwick Township, 4 cm. Reading Public Museum collection 2004C-002-2182.

Figure 256. Chalcopyrite from the French Creek mine, Warwick Township, 18 cm. Carnegie Museum of Natural History collection.

Figure 257. Apophyllite from the French Creek mine, Warwick Township. Field of view is 2 cm. Ron Sloto collection 3371.

Figure 258. Pyrite on magnetite from the French Creek mine, Warwick Township, 3.5 cm. Bryn Mawr College collection.

Figure 259. Magnetite, pseudomorph after hematite, from the French Creek mine, Warwick Township, 5 cm. Ron Sloto collection 1255.

Figure 260. Pyrite on magnetite from the French Creek mine, Warwick Township, 13 cm. Jay Lininger collection.

Figure 261. Apophyllite from the French Creek mine, Warwick Township, 9 cm. Carnegie Museum of Natural History collection.

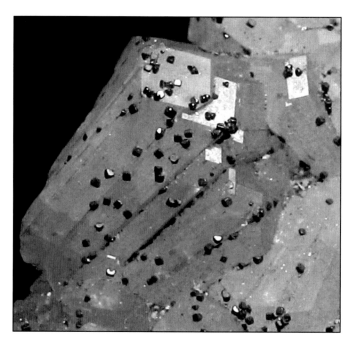

Figure 262. Pyrite on apophyllite from the French Creek mine, Warwick Township, magnified. Ron Sloto collection 2735.

Figure 263. Pyrite on calcite from the French Creek mine, Warwick Township, 4 cm. Ron Sloto collection 3373.

Figure 264. Garnet from the French Creek mine, Warwick Township, 6.8 cm with crystals to 1.5 cm. Bryn Mawr College collection Vaux 5456.

Figure 265. Chromian clinochlore, var. kammererite, from the Scott mine, West Nottingham Township, 3.8 cm. Steve Carter collection.

Figure 266. Chrysotile from the Sparvetta quarry, West Nottingham Township. Field of view is 6 cm. Ron Sloto collection 3077.

Figure 267. Andradite from the Hopewell mine, Warwick Township, 2.5 cm. Bryon Brookmeyer collection.

Figure 268. Andradite from the Hopewell mine, Warwick Township, 1.5 cm. Bryn Mawr College collection Heyl 1350. Collected by Allen Heyl in the 1930s.

Figure 269. Magnetite from the Hopewell mine, Warwick Township, crystals to 5 mm. Bryn Mawr College collection Rand 5322.

Figure 270. Andradite from Knauertown (French Creek mine), Warwick Township, 8.6 cm. Carnegie Museum of Natural History collection CM-6964 (Jefferis 1370).

BRINTON'S QUARRY   WESTTOWN TOWNSHIP

Brinton's quarry produced more serpentine for building stone than any other quarry in the U.S. Quarrying for building stone began around 1720. From 1720 to about 1850, there was no ownership of the quarries; they were considered a "commons" with members of the local community free to take whatever quantity of stone they needed.

Eventually Joseph H. Brinton acquired sole control. Stone from Brinton's quarry was widely used for building and, to a lesser extent, for ornamental stone. Its principal markets were Philadelphia, New York, Washington, D.C., Baltimore, and Chicago. Brinton employed many skilled stone cutters in the quarry to furnish pre-cut stone to architects to reduce the cost of transportation. The architects provided a plan of the building, and each stone was cut in the quarry to fit its appropriate place in the wall. Serpentine fell out of favor as a building stone because it is attacked by the acid-laden atmosphere of cities and rapidly disintegrates. Orders for building stone began to decline in the late 1890s, and it became unprofitable to work the quarry. The quarry was abandoned in 1900.

Brinton's quarry is in a lens of serpentinite surrounded by schist. A white granitic dike several feet wide cuts the serpentine. The serpentinite also is cut by irregular pegmatite dikes. Brinton's quarry is the type locality for clinochlore, which was called ripidolite when first discovered. Clinochlore was named and first described in 1851. Brinton's quarry also is the type locality for the jeffersite variety of vermiculite/hydrobiotite. It was named in honor of West Chester mineral collector William W. Jefferis.

Figure 271. Clinochlore from Brinton's quarry, Westtown Township, 1.6 cm. Bryon Brookmeyer collection.

Figure 272. Beryl from Brinton's quarry, Westtown Township, 5.5 cm. Carnegie Museum of Natural History collection CM-6475 (Jefferis 7009).

Figure 273. Vermiculite/hydrobiotite, var. jeffersite, from Brinton's quarry, Westtown Township, 6.7 cm. Ron Sloto collection 2951.

Figure 274. Clinochlore from Brinton's quarry, Westtown Township, 5.1 cm. Bryon Brookmeyer collection.

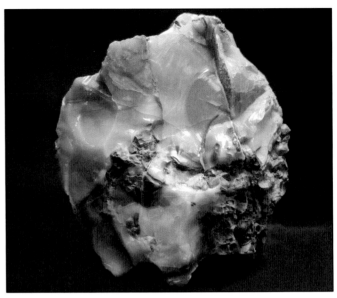

Figure 275. "Deweylite" from Brinton's quarry, Westtown Township, 11.4 cm. West Chester University collection.

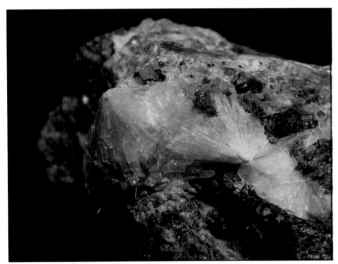

Figure 276. Aragonite from Brinton's quarry, Westtown Township. West Chester University collection.

Figure 277. Chromian clinochlore from Brinton's quarry, Westtown Township, 7 cm. Bryn Mawr College collection Vaux 6905. Originally sold by the A.E. Foote Mineral Company.

Figure 278. Quartz, var. amethyst, from Brinton's quarry, Westtown Township, 3.8 cm. Bryn Mawr College collection Rand 4867. Collected in 1891.

Figure 279. "Deweylite" from Brinton's quarry, Westtown Township, 11.4 cm. West Chester University collection.

Figure 280. Clinochlore from Brinton's quarry, Westtown Township, 10 cm. Steve Carter collection.

Figure 281. Clinochlore from Brinton's quarry, Westtown Township, 5.1 cm. Steve Carter collection.

Figure 282. Schorl from Brinton's quarry, Westtown Township. Field of view 9 is cm. Ron Sloto collection 2961.

Figure 283. Clinochlore, var. roseite, from Brinton's quarry, Westtown Township, 4.4 cm. Ron Sloto collection 2631. Originally sold by the A.E. Foote Mineral Company.

Figure 284. Titanite from the Osborne Hill mine, Westtown Township. Bryn Mawr College collection Vaux 7045.

Figure 285. Zoisite from Bath Springs (West Chester Water Works), West Chester, 9 cm. Carnegie Museum of Natural History collection CM-7559 (Jefferis 94).

Figure 286. Quartz, var. chalcedony, from Willistown Township, 8.5 cm. Delaware County Institute of Science collection. Collected about 1878.

Figure 287. Spessartine from the Osborne Hill mine, Westtown Township, 7 cm. Carnegie Museum of Natural History collection CM-6836 (Jefferis 6906).

Figure 288. Talc from West Goshen Township, 8.5 cm. Bryn Mawr College collection Rand 3017.